VICTORIAN LIFE

A VICTORIAN FACTORY

LYN GASH & SHEILA WATSON

Wayland

VICTORIAN LIFE

HOW WE LEARN ABOUT THE VICTORIANS

Queen Victoria reigned from 1837 to 1901, a time when Britain went through enormous social and industrial changes. We can learn about Victorians in various ways. We can still see many of their buildings standing today, we can look at their documents, maps and artefacts – many of which can be found in museums. Photography, invented during Victoria's reign, gives us a good picture of life in Victorian Britain. In this book you will see what Victorian life was like through some of this historical evidence.

Series design: Pardoe Blacker Ltd
Editor: Sarah Doughty

First published in 1993 by Wayland (Publishers) Ltd
61 Western Road, Hove, East Sussex BN3 1JD, England

British Library Cataloguing in Publication Data
Gash, Lyn
Victorian Factory. - (Victorian Life Series)
I. Title II. Watson, Sheila III. Series
338.60941

ISBN 0 7502 0794 9

Printed and bound in Great Britain by B.P.C.C Paulton Books Ltd

Cover picture: Workers in a Victorian factory.

Picture acknowledgements
Archiv für Kunst und Geschichte, Berlin 5 (top); Blackburn Museum and Art Gallery 11 (top); Cadbury Ltd 19; Colman's of Norwich 11 (bottom), 17; Mary Evans 5 (bottom), 6 (bottom), 13 (top), 15, 17 (bottom), 21, 24, 26, 27 (top); E.T Archive *cover;* Fotomas Index 12; P. Hayden 25 (top); Hulton-Deutsch Collection 22; J.& P. Coats (U.K) Ltd 18 (top); Billie Love Historical Collection 8, 9 (top), 20; Manchester City Art Galleries 10; Mansell Collection 16, 23; National Railway Museum 27 (bottom); National Trust (Mike Williams) 4; Norfolk Museums Service (G.G.S Photographers) 25 (bottom); Punch Picture Library 18 (bottom); Ann Ronan Picture Library 6 (top), 14 (bottom); Salford Working Class Movement Library 9 (bottom); Tony Stone Worldwide (Tom Sheppard) 7; The Wellcome Institute Library, London 13 (bottom).

CONTENTS

Changing Technology

There were factories in Britain before Victoria became queen in 1837. Machinery in factories could do the jobs of many people working by hand. When Victoria's reign ended in 1901, Britain's factories had helped to make it the richest country in the world. But not all goods produced in Victorian times were made in factories. Many people still worked in cottage industries, making goods at home.

THE FIRST FACTORIES

The first water-powered spinning factory was built in 1771 by Richard Arkwright, at Cromford in Derbyshire. He had invented a machine that could make cotton thread much faster than it could be made by hand.

His machine needed water to make it work. The water turned a wheel which powered the factory machinery. Most of the first factories in the 1700s were built in the countryside and driven by water power from fast-flowing streams.

Quarry Bank Mill in Styal was driven by water power.

STEAM ENGINES

In 1781, James Watt, a Scotsman, invented a method of using steam engines to drive factory machines. His invention worked by rotary power. A series of cogs and wheels driven by steam were used to turn the driving belt on a factory machine.

The rotary engines invented by Watt were used in all sorts of industries, especially for manufacturing cotton and textiles.

Steam engine in factory, 1860.

FACTORY TOWNS

The steam-powered Victorian factories were usually built in towns. The factory engines needed fuel to heat up water to make steam. Local coalfields provided the factories with coal to burn as fuel. Towns like Leeds became very smoky because of the many factories that were built during the Victorian period.
These industrial towns grew very quickly. For example, Manchester's population was 75,000 in 1801. By 1851, this had grown to 303,000.

Smoke pollution in Leeds in 1885.

MACHINE SPINNING

Factories like the one shown in the picture on the right could do the work of hundreds of people. This cotton mill required only a few workers to keep the machines going. The items produced were made much faster and more cheaply by machine than by hand. Factory owners usually made plenty of money, and many more people were able to buy these new, cheap products. The cotton industry was one of the most important industries in Victorian Britain.

Spinning cotton by machine, London, 1870.

A dressmakers' workroom in 1858.

SWEATSHOPS

Although factories were growing, many people still worked at home or in sweatshops like the one above. The picture shows women doing fine sewing which could not be done by machines. Can you see how tired and ill some of them look? They had to work long hours in small, crowded rooms for very little money. Sometimes the goods they made were sold to factories.

MODERN FACTORIES

It was the Victorians who developed factories and made them so successful. Today most of our goods are made or processed in factories by machines. In the last thirty years, some factories have been transformed by the micro-chip revolution. Few people are needed to look after a large factory. Often only one person is required to control a whole area, using computers and video cameras.

Computer-controlled machinery in a modern factory.

FACTORY ROUTINE

Factories changed the way people worked. Factory workers had to start work at fixed times and work at the speed of the machines. They were surrounded by other workers, but they were not allowed to waste time by talking to each other. Many people hated this factory discipline and preferred to work at home, even if this meant earning less money.

COTTAGE WORKERS

This picture shows women working at home, spinning thread. These women could chat to each other as they worked and take a break whenever they wanted. They were free to organise their own work routine and there was no one standing over them telling them what to do. They could also take a pride in the way their own hands produced the thread.

Once factories started producing goods cheaply, homeworkers were paid very little for the goods they produced. This forced them to work long hours so they had enough to live on. Some people lost their work altogether and had to find a job in one of the new factories.

Hand-spinners in Scotland.

FACTORY WORKERS

This photograph shows women working in a Victorian factory. Unlike people who worked at home, most factory workers did not make anything with their hands. Their work was to make sure the machines kept running. Factory workers were not allowed to talk to each other or take breaks when they wanted.

In the factories, it was the factory owners not the workers who decided when people should start and finish their work. They wanted to keep their machines going all the time to make as much money as possible. This meant that some workers had to work at night and sleep during the day.

Women at work in a textile factory.

FACTORY RULES

This is a poster that would have been displayed inside a factory. The rules in a factory were harsh and strict. Can you imagine being fined by your teacher for arriving at school dirty or late? Factory workers could have money taken away from them for a large number of things – opening a window without permission, breaking a tool accidentally, or even whistling!

RULES
TO BE OBSERVED
By the Hands Employed in
THIS MILL.

RULE 1. All the Overlookers shall be on the premises first and last.
2. Any Person coming too late shall be fined as follows:—for 5 minutes 2d, 10 minutes 4d, and 15 minutes 6d, &c.
3. For any Bobbins found on the floor 1d for each Bobbin.
4. For single Drawing, Slubbing, or Roving 2d for each single end.
5. For Waste on the floor 2d.
6. For any Oil wasted or spilled on the floor 2d each offence, besides paying for the value of the Oil.
7. For any broken Bobbins, they shall be paid for according to their value, and if there is any difficulty in ascertaining the guilty party, the same shall be paid for by the whole using such Bobbins.
8. Any person neglecting to Oil at the proper times shall be fined 2d.
9. Any person leaving their Work and found Talking with any of the other workpeople shall be fined 2d for each offence.
10. For every Oath or insolent language, 3d for the first offence, and if repeated they shall be dismissed.
11. The Machinery shall be swept and cleaned down every meal time.
12. All persons in our employ shall serve Four Weeks' Notice before leaving their employ; but L. WHITAKER & SONS, shall and will turn any person off without notice being given.
13. If two persons are known to be in one Necessary together they shall be fined 3d each; and if any Man or Boy go into the Women's Necessary he shall be instantly dismissed.
14. Any person wilfully or negligently breaking the Machinery, damaging the Brushes, making too much Waste, &c., they shall pay for the same to its full value.
15. Any person hanging anything on the Gas Pendants will be fined 2d.
16. The Masters would recommend that all their workpeople Wash themselves every morning, but they shall Wash themselves at least twice every week, Monday Morning and Thursday morning; and any found not washed will be fined 3d for each offence.
17. The Grinders, Drawers, Slubbers and Rovers shall sweep at least eight times in the day as follows, in the Morning at 7½, 9½, 11 and 12; and in the Afternoon at 1½, 2½, 3½, 4½ and 5½ o'clock; and to notice the Board hung up, when the black side is turned that is the time to sweep, and only quarter of an hour will be allowed for sweeping. The Spinners shall sweep as follows, in the Morning at 7½, 10 and 12; in the Afternoon at 3 and 5½ o'clock. Any neglecting to sweep at the time will be fined 2d for each offence.
18. Any persons found Smoking on the premises will be instantly dismissed.
19. Any person found away from their usual place of work, except for necessary purposes, or Talking with any one out of their own Alley will be fined 2d for each offence.
20. Any person bringing dirty Bobbins will be fined 1d for each Bobbin.
21. Any person wilfully damaging this Notice will be dismissed.

A list of fines in a factory.

The Dinner Hour, Wigan, a painting showing factory workers at lunch time.

WORKING HOURS

In early factories the working day was very long. It was common to work for twelve hours a day with only a break of fifteen minutes for breakfast and tea and half an hour for the midday meal. Some factories allowed an hour for lunch. Only a few factories had dinner halls or rest rooms.

In the above picture, the women spend their lunch break just outside the factory. The artist has made it look like a happy scene, but imagine how the workers would feel if it was cold or wet – especially the woman without any shoes!

CHEAP LABOUR

Women were employed to look after factory machines but they were only paid half a man's wage. Children were paid even less. Factory owners preferred to employ women and child labour because they were cheaper and so they could get more workers for the same money.

Early Victorian factory owners found even cheaper labour by using orphan apprentices who were given no wages at all. They worked in return for poor food and a dirty bed. Do you think it was right for women and children to receive so much less than men for doing the same work?

Girls in a weaving shed in Blackburn.

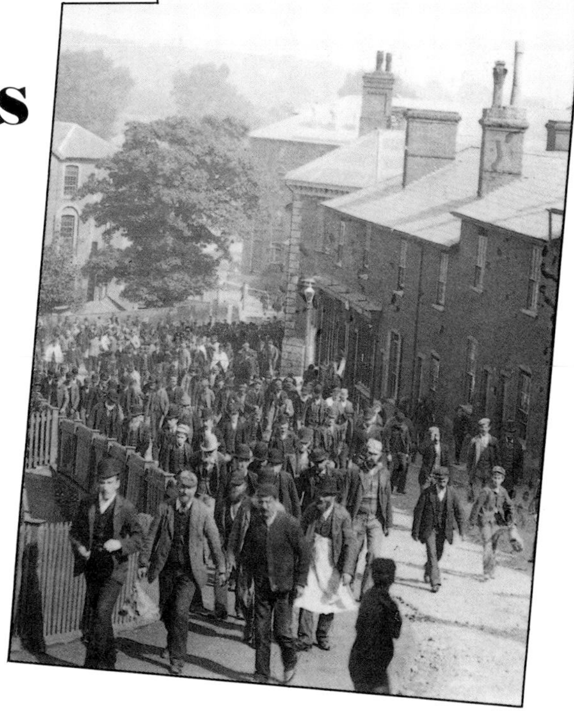

Workers returning home from Colman's factory, Norwich.

AFTER WORKING HOURS

In the early Victorian period there was very little time for anything but work. People usually worked six days a week and the only holidays were Good Friday and Christmas Day. Most workers simply went home to eat and sleep.

Things improved in the late Victorian period as working hours were reduced. The people in this photograph on the right would have been able to meet friends or spend more time with their families later in the evening.

Early factory work was unhealthy and dangerous, as were many other Victorian jobs. The only concern of most factory owners was making money. They did not worry about the health and safety of their workers. There were always many people willing to work in the factories, even if the conditions were poor. Until laws were made to protect the workers, the owners were not forced to think about safety.

A child cleaning a working machine.

SAFETY AT WORK

Can you see the child under the machinery in this picture? Children had to clean fluff away from the machinery while it was still running. This could be very dangerous. Imagine what would happen if workers caught their hair or clothes in the machinery while they were working or leaning over it.

In 1884 a law was passed to force cotton mill owners to put guards over the moving parts of some of their machinery, but it was not until much later that all factories had to protect their workers in this way.

PUNISHMENT

Children were often so tired from working long hours that they could not work quickly or concentrate. Men in charge of the children kept them at work by shouting at them or punishing them. Look at what is happening in the picture on the left. A beating like this was quite a common event in some factories. A few overseers were very cruel. For example a common punishment in one nail-making factory in the Midlands was to drive a nail through the worker's ear into the wooden bench. There were no laws to protect children against such cruelty.

A factory overseer beating a child worker.

CHEMICAL DANGER

Workers often had to use dangerous chemicals in their work. For example, phosphorous was used to make matches. Its fumes could could damage a matchworker's bones. You can see from this picture of a skull how the teeth have dropped out and even the jaws have rotted away due to phosphorous. Until the later part of the Victorian period, there were no laws to punish employers for unhealthy working conditions.

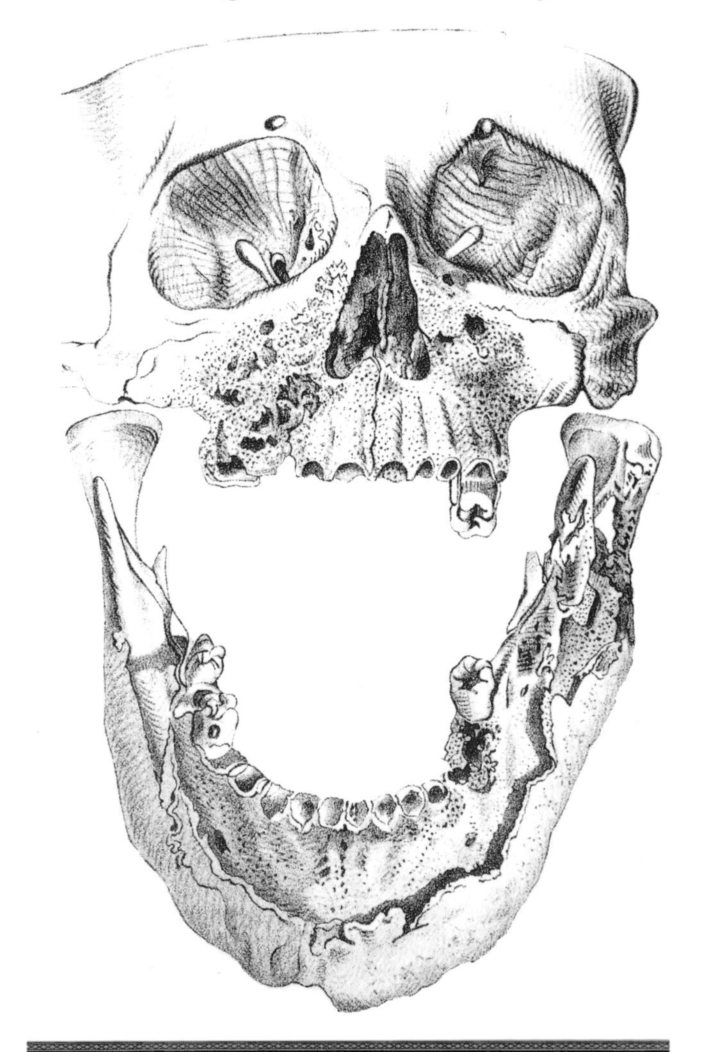

Phosphorous could destroy the matchworker's jawbone.

WORKING CONDITIONS

Factories could be very crowded and very noisy. There were no rules about how many people could work in one factory, or whether there was too much noise from the factory machinery.

In a Victorian factory, workers would spend hours in a noisy, crowded room with poor lighting. The atmosphere could be stuffy in summer or cold with no heating in winter. The women in this photograph look exhausted. Just imagine what the noise would be like with all those tightly-packed machines working in the same room.

Women workers in an ammunition factory in Dundee, Scotland.

INDUSTRIAL ILLNESSES

If people became ill at work, their family or friends took them home. There were no doctors or nurses working, even in the largest factories. Workers were given no money as compensation by the factory owners if their illness or injury was caused by bad working conditions. In fact, workers who missed a day's work, missed a day's pay. Workers who could no longer work ended up in the workhouse, unless their family could support them and look after them. It was only after 1895 that some workers were given compensation for industrial illness.

A worker is taken ill in a factory.

HOMEWORKERS

Conditions might have been bad in factories, but they were no better if you worked at home. This picture shows a family working together. Even the young children are included. The pay for homeworkers was so small that they had to work for long hours in dingy, cold rooms. Some people even did dangerous things like making matches at home.

Other people worked at home on their own. For those people, work could be very lonely. At least if you worked in a factory, you had the companionship of the other workers.

Matchbox workers at home.

Factory conditions gradually improved in the Victorian period. Some factory owners cared about their workers and set a good example to others. Other individuals were shocked by workers' conditions and campaigned to have them improved. Throughout Queen Victoria's reign, many new laws that helped factory workers were passed.

Buildings surround the factories of New Lanark.

NEW LANARK, SCOTLAND

Robert Owen was the owner of cotton mills in New Lanark in Scotland. He tried to improve the working conditions of the people who worked for him. He built schools and villages around his factories so his workers could live in healthy surroundings. He cut the working hours of his employees and stopped using children under ten years of age.

Robert Owen believed that owners could treat their workers well and still get good profits from the factories. However, few other factory owners agreed with him.

COLMAN'S FACTORY SCHOOL

Most factory owners believed they should be allowed to run their factories as they chose. It was only the stories of helpless children being badly-treated that made Parliament pass laws to protect the young workers. The first factory laws tried to keep children's working hours to less than twelve hours a day and give them a basic education.

This picture shows Carrow School, set up by John Colman in 1857 for his factory in Norwich. It was one of the first schools to be set up for the children of factory workers.

The kindergarten class of Carrow school, Norwich, 1890.

FACTORY INSPECTORS

The first factory laws did not work well because there was no one to make sure they were being carried out. In 1833 a new law created the job of factory inspector. However, there were only four inspectors for all the factories in the country. This picture shows children being questioned about their age and working conditions by a factory inspector.

It is easy to see how factory owners managed to avoid the law. Owners and overseers could lie about the children's age and tell the children what to say. See how the factory overseer at the back of this picture is listening to everything that is said.

A government inspector visiting a factory.

HOURS AND CONDITIONS

Gradually more inspectors were appointed and factory owners followed the law. A law passed in 1833 had stopped children under the age of nine working in textile mills. Later laws protected children and women in other types of factories by shortening their working hours.

The Victorian factory laws did not shorten the hours men spent working. However, without the help of women and children, men often could not carry on their own work, so their hours became shorter too.

Children working at Coat's thread factory.

RICH AND POOR

A cartoon comparing a rich lady and poor dressmaker.

PIN MONEY.

NEEDLE MONEY.

The artist of the cartoon on page 18 is comparing the lives of two women. The rich lady in the expensive ballgown on the left would have had plenty of money to spend on herself. The woman who made her dress on the right would have worked long hours for very little pay. The laws that improved life for the factory workers did nothing to help the homeworkers. By the end of the Victorian period, it was the factory workers who had better working conditions and pay than many other people who worked at home.

Cadbury's factory outing, 1880s.

HOLIDAYS

By the end of the Victorian period some factories allowed their workers to take holidays. Some, as in the picture above, even had an annual outing. They hired a train or bus and went to the local seaside resort or picnic spot. It was still unusual for workers to have paid holidays, but some factory owners had begun to understand that people worked better if they were healthy and happy.

Throughout the Victorian period, some factory workers tried to help themselves. They joined trade unions, Friendly Societies and tried to become better educated. Through their efforts they hoped to make the factory owners change the way they treated their workers. Trade unions were hated and feared by many factory owners and they tried to destroy them. However, gradually the government gave trade unions more and more power.

Four of the Tolpuddle martyrs.

THE TOLPUDDLE MARTYRS

The four men in the picture are farmworkers from Tolpuddle in Dorset. In 1834, they (and two others) were transported to Australia and forbidden to return for seven years. This was because they asked new trade union members to take a secret oath, which at that time was forbidden by law.

Employers had a great deal of influence on the people who made the laws. They feared trade unions because they thought the unions would force them to pay higher wages. Employers hoped the punishment for these men would frighten people so much that they would not join trade unions.

EDUCATION CLASSES

Many working people had a very poor education. To try and improve their knowledge they went to lessons and lectures in their spare time. Workers could attend the Working Men's Institutes (schools for working adults), or go to Bible classes in church or chapel, or evening classes. They hoped that better education would improve their working lives.

An evening class for working men.

An advertisement for a Friendly Society.

YOU SHOULD JOIN
THE BEST APPROVED AND REGISTERED SOCIETY UNDER THE ACT.
THE ORDER OF THE SONS OF TEMPERANCE
LOVE
PURITY
FIDELITY
FRIENDLY SOCIETY
Unsurpassed Sick and Assurance Benefits for MEN. WOMEN & CHILDREN.
APPLY TO WRITER.
Registered Copyright.

FRIENDLY SOCIETIES

If Victorian workers became sick or lost their job they could not expect any money from the government to help them. Instead, once their savings had been used up, they starved or went to the workhouse. Better-paid workers tried to avoid this by saving a small amount every week with a Friendly Society. If they became ill, unemployed or needed money for a family funeral, the Friendly Society would help them. The Society could only help the few workers who could afford to make the weekly payments.

MATCHMAKERS' STRIKE

During the 1850s-70s, large numbers of trades unions were formed and were given more power to help workers. In 1875, trade union members were allowed to picket peacefully. The London matchgirls' strike of 1888 was one of the first successful strikes by poorly-paid workers. Low wages and industrial illness caused by bad working conditions led to a successful strike which greatly improved the women's pay and conditions.

Matchmakers striking for better pay and conditions.

A trade unions' meeting of protest, 1834.

TRADE UNIONS

By the end of Victoria's reign, trade unions were much more powerful than they had been in 1834 when this demonstration was held. It was partly due to the efforts of the trade unions that some factory workers' pay and conditions had improved by 1901. Compared with today, though, most people still worked very long hours for low wages.

By the end of Victoria's reign, factories were producing more and more goods for the home. As demand grew, goods became cheaper. At the same time, some workers' wages were rising, so they could buy more of these goods. The middle-classes filled their homes with factory-made items. However, some poor people could not afford these things, even when they were very cheap.

A POOR WORKER'S HOME

This picture from 1862 shows how an artist sees the home of a family of poor workers. Factories were producing plenty of cheap household items but the family that lived in this home could not afford to buy many of them. The room is almost bare. Can you see anything in the picture which was probably made in a factory?

An artist's drawing of a cotton worker's home.

MASS PRODUCTION

At the beginning of Victoria's reign, factory owners began to look for more products to make in factories. They started to mass-produce items such as mangles for wringing out the washing. People were keen to buy gadgets that helped them with everyday tasks such as cleaning. Other labour-saving devices for the home also became available. Factory goods were beginning to make peoples' everyday lives easier.

Mangles were mass-produced in factories.

HOME COMFORTS

This picture shows a room that has been made to look like a Victorian fisherman's home. The cast iron range (oven), the large pots and pans, the barometer on the wall, much of the pottery, the net curtains, the wallpaper, and the fringe round the fireplace are all factory-made. These are the type of goods some working-class people could afford at the end of Victoria's reign. People with a little money could now afford a few luxuries as well as everyday items.

A reconstruction of a fisherman's home, Cromer.

THE GREAT EXHIBITION

In 1851, the Great Exhibition was held in London. There were thousands of factory-produced items and working models of machinery on show. Many of the goods were luxury items which the manufacturers had invented. Others were basic items, which the factories could supply to buyers in quantity. The Exhibition was a great success. Over six million people from all over the world came to visit the Exhibition which was open for 141 days.

The machinery court of the Great Exhibition, 1851.

TRADE CATALOGUE

This is a trade catalogue from a large store.The goods from these types of catalogues would be sent all over the world, especially to parts of the British Empire such as India, South Africa, Canada, Australia and New Zealand.

British goods were bought by people all over the world. In turn, Britain bought raw materials such as wool and cotton, and foodstuffs such as tea, wheat and beef from other countries.

A trade catalogue from a large store.

FLETCHER, RUSSELL & CO., LIMITED,

MANUFACTURERS OF

Gas Cookers, Gas Fires & Heating Stoves,
Water Heaters, Grillers, Washers,
Dryers, Smoothing Iron Heaters,
Internally Heated Smoothing Irons,
and all kinds of Domestic
Labour Saving Appliances.

WARRINGTON, MANCHESTER, AND LONDON.

Showrooms: 134, Queen Victoria St., London, E.C.
130, Deansgate, Manchester.

PROGRESS

Britain produced heavy industrial goods such as iron girders, railway engines, glass and building materials. Factory production meant that these items could be made much more cheaply, so earlier inventions like the railways expanded greatly in Victorian times.

Mass production in factories changed British life forever. Some people think that because of all the factory-produced goods, life must have become better for the Victorians. Others look at the conditions in the factories and think that life had become worse. What do you think?

A railway station. Heavy industrial goods like engines were produced in factories.

TIME LINE

BC	AD 0	43	410	500	
CELTS		ROMAN BRITAIN	'THE DARK AGES' ANGLO-SAXONS		VIKINGS

PRE-1830s

1771 The first water-powered spinning factory built by Richard Arkwright in Cromford, near Derby.

1780s James Watt's invention, the steam-engine, was used to power factory machines.

1799 and **1800** Laws were passed that did not allow people to form trade unions or to go on strike.

1802 Textile Apprentices Act passed to protect orphan apprentices in cotton and woollen factories.

1819 Cotton Factories Act passed to protect orphan apprentices in cotton factories.

1824 Earlier laws were changed so workers could form trade unions, but they had little power.

1830s

1833 Textile Factory Act passed. No children under nine years old allowed to work. Children aged nine to thirteen years old only to work nine hours per day. Every child to have two hours teaching per day. Four factory inspectors appointed to check up on factories.

1834 Grand National Consolidated Trades Union formed. Many trade unions banded together to try to increase their power.

Tolpuddle martyrs. Six farm labourers were transported for seven years for asking new members of a union to take a secret oath. They were later pardoned and allowed to return home.

1840s

1844 Textile Factory Act. Women were not allowed to work more than a twelve hour day.

Dangerous machinery to have guards fitted for safety.

1847 Ten Hour Act. A ten hour day for all women and young people under the age of eighteen.

1850s

1850 Factory Act. Machinery only allowed to operate between the hours of 6 am and 6 pm. But also the working hours for women and children were increased to ten and a half per day.

1851 The Great Exhibition. The first international trade exhibition held in Hyde Park, London.

Amalgamated Society of Engineers was formed which was one of the first of the new model unions. Only skilled workers became members.

1000		1500				2000
1066 NORMANS	MIDDLE AGES	1485 TUDORS	1603 STUARTS	1714 GEORGIANS	1837 VICTORIANS	1901 20TH CENTURY

1860s

1867 Factory Extension Act. Factory law now covered all workplaces which employed more than fifty workers.

Master and Servant Act. Workers could not be dismissed without notice from their employers.

1870s

1871 Bank Holiday Act. Banks closed for a few days a year. This gave many other workers a few extra days holiday too.

Trade Union Act. Trade union funds protected by law.

Criminal Law Amendment Act passed which did not allow peaceful picketing.

1875 Conspiracy and Protection of Property Act. This allowed trade unions to use peaceful picketing during strikes.

1880s

1888 London matchgirls' strike.

1889 London dockers' strike.

1889 London gas workers' strike.

1890s

1893 The Independent Labour Party was formed. This would give workers their own voice in Parliament in the future.

1900s

1901 Minimum working age raised to twelve years.

Death of Queen Victoria.

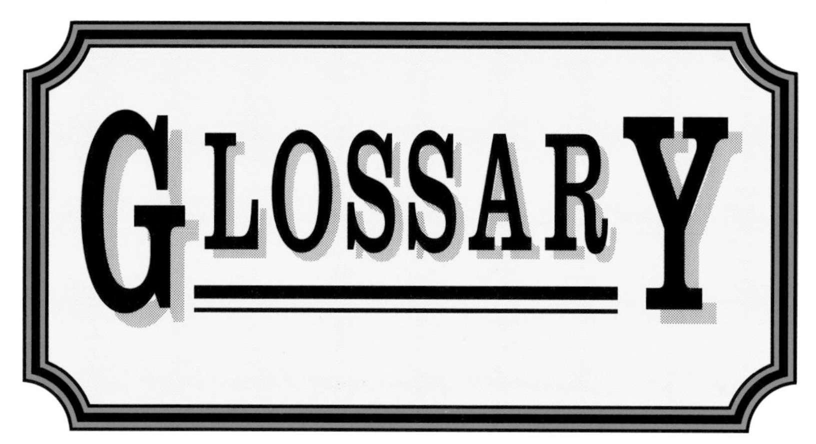

Apprentice A young person that is trained to do a job and forced by law to work at that job for a fixed length of time.
British Empire The countries that were under British control, which included Canada, India, Australia and New Zealand.
Compensation Money or goods given to make up for injury or death at work.
Cottage industry A system of work where people work in their own homes, often using their own tools or equipment.
Discipline Rules which have to be obeyed.
Dolly A hand tool used to mix water in a washing tub.
Friendly Society A club that looked after its members in times of sickness, unemployment and death.
Industrial illness Sickness caused by work.
Labour The people who worked in return for wages. To labour also means 'to work'.
Mangle A machine with two or more rollers for squeezing water out of washed clothes.
Manufacture To make things by hand or machine, usually on a large scale.
Martyr Someone who suffers for what he or she believes in.
New model union A union formed by well-paid workers from the 1850s onwards. This type of union helped its members in times of trouble.
Oath A solemn pledge or promise.
Orphans Children whose parents have died.
Overseer A person who made sure that workers did their jobs properly. Also known as an overlooker.
Pickets Workers who demonstrate outside a place of work during a strike and try to persuade others not to work.
Phosphorous A solid white substance used in making matches.
Rotary engine An engine that powers machines by using cogs and wheels.
Strike Workers agreeing together not to work until their employer meets their demands.
Sweatshop A place of work where hours are long and pay very poor.
Textile Cloth (eg. wool or cotton).
Trade union A club or organization for people in the same kind of work which looks after their pay, hours and safety.
Transported Sent to a prison (or penal colony) abroad, usually Australia.
Wage A payment made in return for work.
Without notice No warning given to someone that they are to lose their job.
Workhouse A place where very poor and homeless people did unpaid work in return for food and a place to sleep. Most people were frightened of the workhouse and thought it was like a prison.

BOOKS TO READ

Evans, D. *Victorians: Early and Late* (A & C Black, 1990)
Huggett, F. *A Day in the Life of a Victorian Factory Worker,* (Allen & Unwin, 1973)
Middleton, G. *The Factory Age* (Longman, 1974)
Ross, S. *A Victorian Factory Worker* (Wayland, 1985)
Ross, S. *Dickens and the Victorians* (Wayland, 1986)
Turner, D. *Victorian Factory Workers* (Wayland, 1990)

PLACES TO VISIT

Many museums have Victorian factory-produced goods and exhibits relating to factory production techniques. A few deal with trade unions and self-improvement. Below is a small selection of all types.

ENGLAND

Cheshire: Quarry Bank Mill, Styal, Cheshire, SK9 4LA. Tel. 0625 527468
Derbyshire: Arkwright's Mill, Cromford, near Matlock. Tel. 062 982 4297
Derby Industrial Museum, The Silk Mill, off Full Street, Derby, DE1 3AR. Tel. 0332 255308
Devon: Coldharbour Mill Working Wool Museum, Uffculme, Cullompton, EX15 3EE. Tel. 0884 840960
Dorset: Tolpuddle Martyrs Museum, TUC Memorial Cottages, Tolpuddle, DT2 7EH. Tel. 0305 848237
Essex: The Braintree and Bocking Heritage Centre, Museum Room, The Town Hall Centre, Market Square, Braintree, CM7 6YG. Tel. 0376 552525, ext 2332
Leeds: Leeds Industrial Museum, Armley Mill, Canal Road, Armley, LS12 2QF. Tel. 0532 637861
London: Science Museum, Exhibition Road, South Kensington, SW7 2DD. Tel. 071 938 8000
Manchester: Manchester Museum of Science and Industry, Liverpool Road, Castlefield, M3 4JP. Tel. 061 832 2244
Rochdale Pioneers Memorial Museum, 31 Toad Lane, Rochdale, Greater Manchester, OL12 ONU. Tel. 061 832 4300
Merseyside: Merseyside Museum of Labour History, Islington, Liverpool L3 8EE. Tel. 051 207 0001
Norfolk: Bridewell Museum of Norwich Trades and Industries, Bridewell Alley, Norwich, NR2 1AQ. Tel. 0603 667228
Nottinghamshire: Ruddington Framework Knitters' Museum, Chapel Street, Ruddington, NR11 6HE. Tel. 0602 846914
Staffordshire: Brindley Mill and James Brindley Museum, Mill Street, Leek, ST13 5PG. Tel. 0538 381446
Shropshire: Ironbridge Gorge Museum Trust, Ironbridge, Telford, TF8 7AW. Tel. 0952 433522
Wiltshire: Trowbridge Museum, The Shires, Court Street, Trowbridge, BA14 8AT. Tel. 0225 751339
Yorkshire: Bradford Industrial Museum, Moorside Road, Eccleshill, Bradford, BD2 3HP. Tel. 0274 631756

SCOTLAND

Glasgow: Tenement House, 145 Buccleuch Street, G3 6QN. Tel. 041 333 0183
New Lanark: New Lanark Conservation Village, New Lanark Mills, Lanark, ML11 9DB. Tel. 0555 61345

WALES

Swansea: Maritime and Industrial Museum, Museum Square, Maritime Quarter, SA1 1SN. Tel. 0792 470371

NORTHERN IRELAND

Belfast: Ulster Folk and Transport Museum, Witham Street Gallery, BT4 1HP. Tel. 0232 451519